NIM
Programming

An Illustrative Guide to
Building a Robust, and Scalable
Applications with Nim

UCHENNA IHEKAIRE

Printed in the United States of America

© 2024

Printed in the

USA

Copyrights

All rights reserved. No part of this publication may be reproduced, stored in a retravel system or transmitted in any form or by any means, electronic, mechanical, photocopying, recording, and scanning without permission in writing by the author.

Printed in the United States of America

TABLE OF CONTENT

PART ONE 1

    THE FIRST PROGRAM 2

    LEXICAL ELEMENTS 7

    THE VAR STATEMENT 11

    CONSTANTS 12

    LET STATEMENT 14

    THE ASSIGNMENT STATEMENT 16

    CONTROL FLOW STATEMENT 18

    THE WHEN STATEMENT 23

    STATEMENTS AND INDENTATION 24

    PROCEDURES 25

    ITERATORS 40

    BOOLEANS 43

    INTERNAL TYPE REPRESENTATION 49

    ADVANCED TYPES 50

PART TWO 89

    PROGRAMS 90

    OBJECT-ORIENTED PROGRAMMING 90

    TEMPLATES 114

    COMPILATION TO JAVASCRIPT 120

PART THREE 121

MACRO ARGUMENTS 121
THE SYNTAX TREE 124
GENERATING CODE 126

INDEX 135

PART ONE

The First Program

Let's dive into this tutorial by exploring the ever-famous "*hello world*" program. It serves as the perfect starting point for beginners in the world of programming.

Nim Codes

```nim
# This is a comment
echo "What's your name? "
var name: string = readLine(stdin)
echo "Hi, ", name, "!"
```

Go ahead and save the file as "greetings.nim", then compile and run the file.

Nim Codes

```
nim compile --run greetings.nim
```

When you make use of the "--run" switch in Nim, something magical happens. After your code gets compiled, Nim takes the initiative to run the program automatically. It saves you the hassle of manually executing the file separately.

Now, let's talk about passing command-line arguments to your program. It's quite simple. Just append the arguments you want to pass right after the filename. This way, your program can effortlessly receive and process the necessary input from the command line.

Nim Codes

```
nim compile --run greetings.nim arg1 arg2
```

When it comes to frequently used commands and switches, abbreviations can be a lifesaver. They allow you to save time and effort by using shorter forms of those commands and switches. This way, you can maintain convenience while working with your preferred programming tools. Don't hesitate to leverage abbreviations whenever you find yourself using certain commands or switches repeatedly. It's all about making your coding experience more efficient and hassle-free.

Nim Codes

```
nim c -r greetings.nim
```

If you're looking to debug your code and identify any issues, this specific version is tailored just for that purpose. However, when it's time to compile a release version of your program, there's a different command you should use. Here's what you need to do:

Nim Codes

```
nim c -d:release greetings.nim
```

By default, the Nim compiler comes equipped with an array of runtime checks that contribute to an enhanced debugging experience. However, if you're aiming to optimize your code and disable certain checks, you can make use of the "-d:release" switch. This switch not only disables specific checks but also enables optimizations that can boost the performance of your code.

When it comes to benchmarking or working with production code, it is highly recommended to utilize the "-d:release" switch. It ensures that your code runs smoothly and efficiently in those scenarios. Additionally, if you wish to compare Nim's performance with other low-level languages like C, you can leverage the "-d:danger" switch. This allows you to obtain meaningful and comparable results, as it disables certain checks that may not be applicable in C.

Now, let's take a closer look at the program's functionality and syntax. The statements that are not indented are executed right at the beginning when the program starts. In Nim, indentation is vital as it groups statements together. It's important to note that only spaces are permitted for indentation, while tabulators should be avoided.

In this specific code snippet, string literals are enclosed within double quotes. The "var" statement declares a new variable named "name" of type string. It is assigned the value returned by the "readLine" procedure. Since the compiler already knows that "readLine" returns a string, you can omit the type in the declaration. This feature is referred to as local type inference and allows for a more concise declaration.

To put it simply, the following alternative declaration will work just as well:

Nim Codes

```
var name = readLine(stdin)
```

It's worth noting that Nim primarily relies on local type inference for type recognition. This form of type inference strikes a balance between conciseness and readability, making it the primary approach in Nim.

Now, let's explore the "hello world" program and its various identifiers. The program utilizes built-in identifiers such as "echo," "readLine," and others. These identifiers are already recognized by the compiler and are declared within the system module, which is automatically imported by any other module.

Lexical Elements

Now, let's dive into the lexical elements of Nim. Like other programming languages, Nim consists of different components, including literals, identifiers, keywords, comments, operators, and punctuation marks.

When it comes to string and character literals, Nim follows familiar conventions. String literals are enclosed within double quotes, while character literals are enclosed within single quotes. Special characters can be represented using escape sequences, where a backslash () is used as the escape character. For example, "\n" represents a newline, "\t" represents a tabulator, and so on. Nim also supports raw string literals, denoted by the prefix "r". In raw literals, the backslash character is not treated as an escape character.

Here's an example of a raw string literal: r"C:\program files\nim". This form is particularly useful when dealing with file paths or other scenarios where escape characters should be treated literally.

Nim also provides long-string literals, which are enclosed within three quotes ("""). These literals can span across multiple lines and allow for easy inclusion of line breaks and special characters. Within long-string literals, the backslash character is not interpreted as an escape character. These long-string literals are commonly used for embedding HTML code templates or other situations where preserving the exact formatting is crucial.

Let's move on to comments. In Nim, comments begin with the hash character (#), unless they appear within a string or character literal. Documentation comments, used for documenting code, start with a double hash character (##). They serve as a way to provide additional information or explanations about the code for future reference.

```
# A comment.

var myVariable: int ## a documentation comment
```

Documentation comments play a vital role in Nim as they are treated as tokens and hold specific positions within the input file, making them an integral part of the syntax tree. This unique characteristic of documentation comments makes it easier to develop streamlined documentation generators.

In addition to documentation comments, Nim also supports multiline comments. These comments start with #[and end with]#. They provide a convenient way to include longer comments or explanations that span across multiple lines. Furthermore, Nim allows nesting of multiline comments within each other, which can be useful when organizing and structuring comments in a more hierarchical manner.

Nim Codes

```
#[
You can have any Nim code text commented
out inside this with no indentation restrictions.
    yes("May I ask a pointless question?")
  #[
    Note: these can be nested!!
  ]#
]#
```

Numbers

When it comes to numeric literals in Nim, the syntax follows a familiar pattern seen in other programming languages. However, Nim adds a special touch to enhance readability by allowing underscores within numeric literals. This means you can use underscores to separate digits and make large numbers more human-friendly. For example, you can write 1_000_000 to represent one million. It's a simple yet effective way to improve the clarity of your code.

Nim also supports floating-point literals, which are numbers that include a dot (.), 'e', or 'E'. These literals represent decimal numbers in scientific notation. For instance, 1.0e9 represents one billion.

If you need to work with hexadecimal numbers, Nim allows you to prefix them with "0x". Binary literals can be indicated with "0b", and octal literals with "0o". It's worth noting that a leading zero alone does not create an octal literal.

The Var Statement

In Nim, the var statement is used to declare new variables, both at the local and global level. It serves as a way to introduce a new variable into the program's scope, giving it a name and specifying its type. With the var statement, you can define variables that store various types of data, such as numbers, strings, or custom data structures.

var x, y: int # declares x and y to have the type
`int`

Indentation can be employed after the var keyword to group a section of variables together:

Nim Codes

```
var
  x, y: int
  # a comment can occur here too
  a, b, c: string
```

Constants

In programming, constants are symbols that represent a fixed value. Once a constant is defined, its value remains unchanged throughout the program's execution. Constants are helpful when you have a value that should not be modified or when you want to assign meaning to a specific value.

In Nim, constants are declared using the "const" keyword. When defining a constant, the compiler evaluates the expression associated with it at compile time. This means that the value of a constant must be known and computable during the compilation process.

By using constants, you can assign meaningful names to values that are used repeatedly in your code. This not only improves code readability but also allows for easier maintenance and modification of values in a centralized manner.

Remember, once a constant is defined, its value cannot be changed. It remains constant throughout the execution of the program, providing a reliable and predictable value for your code.

Nim Codes

```nim
const x = "abc" # the constant x contains the string "abc"
```

Indentation can be utilized after the const keyword to group a section of constants together:

```
const
  x = 1
  # a comment can occur here too
  y = 2
  z = y + 5 # computations are possible
```

Let Statement

In Nim, the let statement serves a similar purpose to the var statement. It allows you to declare symbols and associate them with values. However, there is a key distinction between the two.

When you use the let statement, the symbols you declare are considered single assignment variables. This means that once a value is assigned to a let variable, it cannot be changed or reassigned throughout the program's execution. Let variables are meant to hold constant values that remain unchanged.

```
let x = "abc"   # introduces a new variable `x` and
binds a value to it
x = "xyz"       # Illegal: assignment to `x`
```

Comparing let and const

While let variables are single assignment variables, const has a different purpose in Nim. The const keyword is used to enforce compile-time evaluation of an expression and store its value in a data section. In other words, const signifies that the value associated with it is known and can be evaluated during the compilation process.

The distinction between let and const is as follows: let introduces a variable that cannot be reassigned, while const signifies "enforce compile-time evaluation and store it in a data section". Let variables are used for single assignment, while const is used for compile-time evaluation and storage of values.

Understanding the difference between let and const is important for choosing the appropriate approach based on your specific requirements and design considerations.

Nim Codes

```
const input = readLine(stdin) # Error: constant expression expected
```

Nim Codes

```
let input = readLine(stdin)   # works
```

The Assignment Statement

In programming, the assignment statement is used to assign a new value to a variable or a storage location. It allows you to update the content of a variable during the execution of your program.

When you want to assign a value to a variable in Nim, you use the assignment operator (=). This operator signifies that the value on the right side of the operator is assigned to the variable on the left side.

Nim Codes

```
var x = "abc"   # introduces a new variable `x`
                and assigns a value to it
x = "xyz"       # assigns a new value to `x`
```

Overloading the Assignment Operator

In Nim, it is also possible to overload the assignment operator. This means that you can define custom behavior for the assignment operation based on the types involved. By doing so, you can have more control over how values are assigned and processed.

Declaring Multiple Variables with a Single Assignment Statement

One useful feature in Nim is the ability to declare multiple variables with a single assignment statement. When you have multiple variables that need to be initialized with the same value, you can use a single assignment statement to achieve this.

For example, if you want to declare and assign the value 42 to three variables named "a," "b," and "c," you can write: a = b = c = 42. This assigns the value 42 to all three variables in one statement.

By leveraging the assignment statement, you can dynamically update variable values throughout your program, making it a powerful tool for manipulating and processing data.

Nim Codes

```
var x, y = 3  # assigns 3 to the variables `x` and `y`
echo "x ", x # outputs "x 3"
echo "y ", y # outputs "y 3"
x = 42        # changes `x` to 42 without changing `y`
echo "x ", x # outputs "x 42"
echo "y ", y # outputs "y 3"
```

Control Flow Statement

The Greetings program consists of three statements that are executed in order, one after the other. This sequential execution is suitable for basic programs that follow a straightforward flow. However, as programs become more complex, relying solely on sequential execution is often insufficient. To achieve more advanced functionality, we need to employ branching and looping constructs.

The If Statement

The if statement is a fundamental branching construct in programming. It allows us to control the flow of execution based on certain conditions. With the if statement, we can create branches in our code, leading to different paths of execution depending on whether a condition is true or false.

The syntax of an if statement in Nim is as follows:

```nim
Nim Codes
if condition:
    # code to execute if the condition is true
else:
    # code to execute if the condition is false
```

The condition is an expression that evaluates to a boolean value (either true or false). If the condition is true, the code block following the if statement is executed. If the condition is false, the code block following the else statement (if present) is executed.

By utilizing if statements, we can introduce decision-making capabilities into our programs, allowing them to adapt and respond to different scenarios. It is a crucial construct for building more complex and dynamic functionality.

Case Statement

Handling Multiple Values

In Nim, the case statement provides a flexible way to handle different values or ranges of values. When using the case statement, we can specify a list of values separated by commas within an "of branch." This allows us to execute specific code when the expression matches any of the specified values.

Support for Various Types

The case statement in Nim is versatile and can handle different types of data, such as integers, ordinal types, and strings. Ordinal types represent values that can be ordered or compared, including integers, characters, and enumerated types.

Value Ranges for Integers and Ordinal Types

When dealing with integers or ordinal types, the case statement in Nim also allows us to specify value ranges. This means we can define a range of values and execute specific code if the expression falls within that range.

For example, we can handle a range of values like this:

Nim Codes

```
case expression:
  of minValue .. maxValue:
    # code to execute if expression is within the specified range
  of anotherValue:
    # code to execute if expression matches anotherValue
  else:
    # code to execute if expression does not match any specified values or ranges
```

By using the case statement in Nim, we can effectively handle different values or ranges of values, making our code more readable, organized, and adaptable to various scenarios. It provides a powerful tool for implementing decision-making logic in our programs.

When statement

Example:

Nim Codes
```nim
when system.hostOS == "windows":
```

```
  echo "running on Windows!"
elif system.hostOS == "linux":
  echo "running on Linux!"
elif system.hostOS == "macosx":
  echo "running on Mac OS X!"
else:
  echo "unknown operating system"
```

The When Statement

The when statement in Nim shares similarities with the if statement, but there are a few important distinctions to note:

- **Constant Expressions:** Each condition in the when statement must be a constant expression. These conditions are evaluated by the compiler during compilation rather than at runtime.
- **Scope of Statements:** Unlike the if statement, the statements within each branch of a when statement do not create a new scope. This means that variables declared within a branch can be accessed outside of that branch as well.

- **Selective Code Generation:** The Nim compiler verifies the semantics of the when statement and generates code only for the statements that correspond to the first condition that evaluates to true. This helps optimize the resulting code by excluding unnecessary code paths.

Benefit of the When Statement

The when statement is particularly useful for composing platform-specific code. It allows you to conditionally include or exclude specific sections of code based on compile-time conditions. This makes it comparable to the #ifdef construct in the C programming language.

Statements and Indentation

Now, let's discuss the guidelines for statements and indentation in Nim. In Nim, a distinction is made between simple statements and complex statements.

- **Simple Statements**: Simple statements, such as assignments, procedure calls, and the return statement, cannot include other statements. They do not require indentation and can stand alone.
- **Complex Statements**: Complex statements, such as if, when, for, and while, can contain other statements. To ensure clarity and maintainability, complex statements should always be indented. Indentation helps to visually separate the code blocks and makes the program structure more evident.

By following these guidelines, you can write clean and organized Nim code, utilizing the power of both simple and complex statements effectively.

Procedures

To incorporate new commands like "echo" and "readLine" in the given examples, the concept of procedures becomes essential. While you may be familiar with terms like methods or functions in other programming languages, Nim distinguishes these concepts. In Nim, procedures are defined using the keyword "proc".

Nim Codes

```nim
proc yes(question: string): bool =
  echo question, " (y/n)"
  while true:
    case readLine(stdin)
    of "y", "Y", "yes", "Yes": return true
    of "n", "N", "no", "No": return false
    else: echo "Please be clear: yes or no"

if yes("Should I delete all your important files?"):
  echo "I'm sorry Dave, I'm afraid I can't do that."
else:
  echo "I think you know what the problem is just as well as I do."
```

Let's consider an example that illustrates a procedure named "yes". This procedure prompts the user with a question and returns true if the user responds with "yes" or something similar, and false if they answer "no" or something similar. By using a return statement, the procedure can exit immediately, including the enclosing while loop.

The syntax "(question: string): bool" indicates that the procedure expects a parameter named "question" of type string and returns a value of type bool. The bool type is a built-in type, representing either true or false. It's important to note that conditions in if or while statements must have a bool type.

Result Variable

When a procedure returns a value, it implicitly has a result variable that represents the return value. Using a return statement without an expression is a shorthand notation for "return result". If there is no return statement at the exit of the procedure, the result value is automatically returned at the end.

Nim Codes

```
proc sumTillNegative(x: varargs[int]): int =
  for i in x:
```

```
  if i < 0:
    return
  result = result + i

echo sumTillNegative() # echoes 0
echo sumTillNegative(3, 4, 5) # echoes 12
echo sumTillNegative(3, 4 , -1 , 6) # echoes 7
```

The result variable is implicitly declared at the beginning of the procedure. Therefore, redeclaring it with 'var result' would create a regular variable with the same name, overshadowing the implicit result variable. The result variable is initialized with the default value of its type. It's worth noting that referential data types are initially set as nil and may require manual initialization.

If a procedure lacks a return statement and does not utilize the special result variable, it will return the value of its last expression. For instance, consider the following procedure:

```nim
proc helloWorld(): string =
  "Hello, World!"
```

This procedure returns the string "Hello, World!".

Parameters

Parameters are immutable within the body of a procedure by default. Their values cannot be modified, which allows the compiler to implement efficient parameter passing. If a mutable variable is required within the procedure, it should be declared using the 'var' keyword within the procedure body. Shadowing the parameter name is allowed and is actually a common programming idiom.

Nim Codes

```nim
proc printSeq(s: seq, nprinted: int = -1) =
  var nprinted = if nprinted == -1: s.len else: min(nprinted, s.len)
  for i in 0 ..< nprinted:
    echo s[i]
```

If the procedure needs to modify the argument for the caller, a var parameter can be used:

Nim Codes

```nim
proc divmod(a, b: int; res, remainder: var int) =
  res = a div b     # integer division
  remainder = a mod b  # integer modulo operation

var
  x, y: int
divmod(8, 5, x, y) # modifies x and y
echo x
echo y
```

In the provided example, the parameters "res" and "*remainder*" are declared as var parameters. Var parameters can be modified by the procedure, and any changes made to them are reflected in the caller's scope. However, it is worth considering that using a tuple as a return value would be a more suitable approach for the given example, rather than relying on var parameters.

Discard statement

When invoking a procedure that returns a value solely for its side effects and the return value is to be disregarded, a discard statement must be used. In Nim, silently discarding a return value is not allowed.

Nim Codes

discard yes("May I ask a pointless question?")

If the called procedure or iterator has been declared with the "discardable" pragma, the return value can be implicitly ignored. This means that the return value can be disregarded without explicitly using a discard statement.

Nim Codes

proc p(x, y: int): int {.discardable.} =
  return x + y

p(3, 4) # now valid

Named arguments

In many cases, a procedure may have multiple parameters, which can make it difficult to determine the order in which the arguments should be provided. This is especially relevant for procedures that involve complex data types. To address this, arguments can be named when invoking a procedure, providing clarity on which argument corresponds to which parameter.

Nim Codes

proc createWindow(x, y, width, height: int; title: string;
        show: bool): Window =
    ...

var w = createWindow(show = true, title = "My Application",
        x = 0, y = 0, height = 600, width = 800)

By using named arguments when calling the "*createWindow*" procedure, the order of the arguments becomes insignificant. This approach allows for flexibility in specifying the arguments regardless of their original order. It is also possible to mix named arguments with ordered arguments, although this can potentially impact code readability.

Nim Codes

```
var w = createWindow(0, 0, title = "My Application",
          height = 600, width = 800, true)
```

The compiler ensures that each parameter receives exactly one argument.

Default values

To improve the usability of the "*createWindow*" procedure, incorporating default values can be beneficial. Default values serve as predefined arguments that are used if the caller does not explicitly provide them.

Nim Codes

```
proc createWindow(x = 0, y = 0, width = 500, height = 700,
```

title = "unknown",
show = true): Window =

...

var w = createWindow(title = "My Application", height = 600, width = 800)

Now, when invoking the "*createWindow*" procedure, it is only necessary to provide values that differ from the default ones.

It's worth noting that type inference applies to parameters with default values, eliminating the need to explicitly specify the type and default value, such as "title: string = *'unknown*".

Overloaded procedures

Nim provides the ability to overload procedures, similar to C++. This means that multiple procedures can share the same name but have different parameter lists or argument types. Overloading offers greater flexibility and convenience when working with procedures that serve similar purposes but operate on different data types or with varying sets of parameters.

Nim Codes
```nim
proc toString(x: int): string =
  result =
    if x < 0: "negative"
    elif x > 0: "positive"
    else: "zero"

proc toString(x: bool): string =
  result =
    if x: "yep"
    else: "nope"

assert toString(13) == "positive"   # calls the toString(x: int) proc
assert toString(true) == "yep"      # calls the toString(x: bool) proc
```

(Keep in mind that in Nim, the "$" operator is commonly used for the "*toString*" functionality.) When calling the "*toString*" procedure, the compiler automatically selects the most appropriate implementation based on the context. The specifics of how the overloading resolution algorithm works are not covered here; for more information, refer to the Nim manual. It's important to note that ambiguous calls will result in reported errors.

Operators

The Nim standard library extensively utilizes overloading, as each operator (e.g., +) is an overloaded procedure. The parser allows operators to be used in infix notation (a + b) or prefix notation (+ a). Infix operators always receive two arguments, while prefix operators only receive one argument. Postfix operators are not allowed to avoid ambiguity. For example, does a @ @ b mean (a) @ (@b) or (a@) @ (b)? In Nim, it always means (a) @ (@b), as there are no postfix operators.

With the exception of a few built-in keyword operators like and, or, not, operators consist of specific characters: + - * \ / < > = @ $ ~ & % ! ? ^ . |

User-defined operators are allowed. There is no restriction on defining your own @!?+~ operator, but it's worth considering that doing so may impact code readability.

The precedence of an operator is determined by its first character. Further details can be found in the Nim manual.

To define a new operator, enclose it in backticks "`".

Nim Codes

proc `$` (x: myDataType): string = ...
now the $ operator also works with myDataType, overloading resolution
ensures that $ works for built-in types just like before

The "`" notation can also be used to call an operator just like any other procedure:

Nim Codes

if \`==\`(\`+\`(3, 4), 7): echo "true"

Forward declarations

In Nim, it is necessary to declare every variable, procedure, and other entities before they can be used. This requirement exists due to the complexity of avoiding such declarations in a language that extensively supports metaprogramming, as Nim does. However, a challenge arises when dealing with mutually recursive procedures as it becomes impossible to declare them in advance.

Nim Codes
```
# forward declaration:
proc even(n: int): bool
```

Nim Codes
```
proc odd(n: int): bool =
  assert(n >= 0) # makes sure we don't run into negative recursion
  if n == 0: false
  else:
```

```
    n == 1 or even(n-1)

proc even(n: int): bool =
  assert(n >= 0) # makes sure we don't run into negative recursion
  if n == 1: false
  else:
    n == 0 or odd(n-1)
```

In this scenario, the "odd" procedure relies on "even" and vice versa. To ensure the proper functioning of the code, "even" needs to be introduced to the compiler before it is fully defined. This can be achieved by omitting the "=" sign and the procedure's body, which serves as a forward declaration. It's worth noting that the "assert" statement included in the example accounts for boundary conditions and will be covered in more detail in the Modules section at a later stage.

It's important to mention that future versions of the language will relax the requirements for forward declarations.

Additionally, the example demonstrates that a procedure's body can consist of a single expression, and the value of that expression is implicitly returned.

Functions and methods

As mentioned earlier, Nim distinguishes between procedures, functions, and methods, each defined by the keywords "proc," "func," and "method," respectively. Nim's definitions in this regard are more meticulous compared to other programming languages.

Functions closely align with the concept of a pure mathematical function, which may be familiar if you have experience with functional programming. Essentially, functions are procedures with additional constraints. They are unable to access global state (except for "const" variables) and cannot produce side effects. The "func" keyword serves as an alias for "proc" and is annotated with "{.noSideEffects.}". However, functions can still modify their mutable arguments, indicated by the "var" keyword, as well as any reference objects.

In contrast to procedures, methods are dynamically dispatched. This concept is closely linked to inheritance and object-oriented programming. When overloading a procedure (which refers to having two procedures with the same name but different types or argument sets), the appropriate procedure to use is determined at compile-time. On the other hand, methods rely on objects that inherit from the "RootObj" class. A more comprehensive exploration of this topic is covered in greater detail in the second part of the tutorial.

Iterators

Let's revisit the simple counting example:

```nim
echo "Counting to ten: "
for i in countup(1, 10):
  echo i
```

Can we write a countup procedure that supports this loop? Let's try:

Nim Codes
```
proc countup(a, b: int): int =
  var res = a
  while res <= b:
    return res
    inc(res)
```

However, this approach is not suitable. The issue lies in the fact that the procedure should not only return but also resume execution after an iteration is completed. This functionality, which involves both returning and continuing, can be achieved through the use of a yield statement. To implement this behavior, all that remains is to replace the "proc" keyword with "iterator." With this modification, we have successfully created our first iterator.

Nim Codes
```
iterator countup(a, b: int): int =
  var res = a
```

```
while res <= b:
  yield res
  inc(res)
```

Iterators share some similarities with procedures, but they have several significant distinctions:

- Iterators can only be invoked within for loops.
- Return statements are not allowed in iterators (yield statements are not permitted in procedures).
- Iterators do not have an implicit result variable.
- Recursion is not supported in iterators.
- Forward declaration of iterators is not possible because the compiler needs the ability to inline them (although this restriction will be lifted in a future version of the compiler).

However, it's important to note that you can utilize a closure iterator to impose a different set of limitations. For more information, refer to the details on first-class iterators. Additionally, iterators can share the same name and parameters as a procedure since they have their own namespaces. As a result, it is common to encapsulate iterators within procedures of the same name. These procedures accumulate the iterator's result and return it as a sequence, as demonstrated by the "split" function in the "strutils" module.

Basic Types

This section offers a comprehensive overview of the fundamental built-in types in Nim and the available operations for each type.

Booleans

Nim's boolean type is called "bool" and consists of two predefined values: "true" and "false." Conditions in "while," "if," "elif," and "when" statements must be of type "bool."

The bool type supports various operators, including "not," "and," "or," "xor," "<," "<=," ">," ">=," "!=", and "==." The "and" and "or" operators perform short-circuit evaluation. For example:

Nim Codes

```
while p != nil and p.name != "xyz":
  # p.name is not evaluated if p == nil
  p = p.next
```

Characters

The character type in Nim is represented by the "char" keyword. It has a fixed size of one byte, allowing it to represent a single byte of a UTF-8 character. This design choice prioritizes efficiency, as UTF-8 handles multibyte characters correctly. Character literals are enclosed in single quotes.

Comparison operations such as "==," "<," "<=," ">," and ">=" can be used with chars. The "$" operator converts a char to a string. Chars cannot be mixed with integers directly. To obtain the ordinal value of a char, the "ord" procedure is used.

Conversely, converting from an integer to a char is done with the "chr" procedure.

Strings

String variables in Nim are mutable, enabling efficient string appending. Nim strings are zero-terminated and contain a length field. The length of a string can be obtained using the built-in "len" procedure, excluding the terminating zero. Accessing the terminating zero will result in an error. The zero termination facilitates converting a Nim string to a cstring without copying.

The assignment operator for strings performs a copy. The "&" operator is used for string concatenation, and the "add" operator can be used for appending to a string.

String comparison in Nim follows lexicographical order, supporting standard comparison operators. By convention, Nim assumes strings to be UTF-8 encoded, although it's not enforced. When reading strings from binary files, they are treated as a sequence of bytes. Indexing a string, such as "s[i]," refers to the i-th character, not the i-th unichar.

A string variable is initialized with an empty string, represented as "".

Integers

Nim provides several built-in integer types, including "int," "int8," "int16," "int32," "int64," "uint," "uint8," "uint16," "uint32," and "uint64."

The default integer type in Nim is "int." Integer literals can be suffixed with a specific type to indicate a non-default integer type:

Integers are commonly used for counting objects stored in memory, with the "int" type having the same size as a pointer.

Standard arithmetic operators such as "+," "-," "*," "div," "mod," "<," "<=," "==," "!=," ">" and ">=" are defined for integers. Bitwise operations are available through the "and," "or," "xor," and "not" operators. Left bit shifting is performed with the "shl" operator, while right shifting is done with the "shr" operator. Bit shifting operators treat their arguments as unsigned. For arithmetic bit shifts, regular multiplication or division can be used.

Unsigned operations in Nim wrap around, preventing over- or underflow errors.

In expressions involving different integer types, lossless automatic type conversion is performed. However, if the conversion would result in information loss, the "RangeDefect" exception is raised (unless the error is detected at compile time).

Floats

Nim provides built-in floating-point types: "float," "*float32*," and "*float64*."

The default float type in Nim is "float," which is always 64-bits in the current implementation.

Float literals can be suffixed with a specific type to indicate a non-default float type:

Nim Codes

```
# This is a comment
```

Floats adhere to the IEEE-754 standard and support common operators such as "+," "-," "*," "/," "<," "<=," "==," "!=," ">" and ">=".

Automatic type conversion is performed in expressions involving different floating-point types. The smaller type is converted to the larger type. However, automatic conversion between integer types and floating-point types does not occur. To perform such conversions, you can use the "toInt" and "toFloat" procedures.

Type Conversion

Numerical type conversion is performed by using the desired type as a function:

Nim Codes

```
var
  x: int32 = 1.int32   # same as calling int32(1)
  y: int8  = int8('a') # 'a' == 97'i8
```

```
z: float = 2.5      # int(2.5) rounds down to 2
sum: int = int(x) + int(y) + int(z) # sum == 100
```

Internal type representation

As mentioned earlier, the built-in "$" operator, known as the stringify operator, allows you to convert basic types into strings for printing to the console using the "echo" procedure. However, for advanced and custom types, you need to explicitly define the "$" operator. Sometimes, you may need to quickly debug the current value of a complex type without writing its "$" operator. In such cases, you can utilize the "repr" procedure, which works with any type, even complex data graphs with cycles. The following example demonstrates the distinction between the "$" and "repr" outputs, even for basic types:

Nim Codes
```
var
  myBool = true
```

```
  myCharacter = 'n'
  myString = "nim"
  myInteger = 42
  myFloat = 3.14
echo myBool, ":", repr(myBool)
# --> true:true
echo myCharacter, ":", repr(myCharacter)
# --> n:'n'
echo myString, ":", repr(myString)
# --> nim:0x10fa8c050"nim"
echo myInteger, ":", repr(myInteger)
# --> 42:42
echo myFloat, ":", repr(myFloat)
# --> 3.14:3.14
```

Advanced Types

In Nim programming, you can define new types within a type statement:

Nim Codes

```
type
  biggestInt = int64     # biggest integer type that is available
  biggestFloat = float64  # biggest float type that
```

is available

Enumeration and object types in Nim can only be defined within a type statement.

Enumerations

Variables of an enumeration type can only be assigned specific values defined within the enumeration. These values form an ordered collection of symbols, where each symbol is internally mapped to an integer value. At runtime, the first symbol is represented by 0, the second symbol by 1, and so on. For example:

Nim Codes
```
type
  Direction = enum
    north, east, south, west

var x = south    # `x` is of type `Direction`; its value is `south`
```

```
echo x        # prints "south"
```

Comparison operators can be used with enumeration types.

To avoid ambiguities, an enumeration symbol can be qualified by specifying the enumeration name followed by a dot, such as "Direction.south."

The "$" operator can convert an enumeration value to its corresponding name, while the "ord" procedure can convert it to its underlying integer value.

For better interoperability with other programming languages, the symbols of enum types can be explicitly assigned ordinal values. However, it is important to ensure that the ordinal values are assigned in ascending order.

Ordinal Types

Ordinal types encompass enumerations, integer types, characters, booleans, and subranges. These types have a well-defined order. Ordinal types come with several special operations:

Operation	Comment
Ord(x)	returns the integer value that is used to represent x's value
Inc(x)	increments x by one
Inc(x, n)	increments x by n; n is an integer
dec(x)	decrements x by one
Dec(x, n)	decrements x by n; n is an integer
Succ (x)	returns the successor of x
Succ(x, n)	returns the n'th successor of x
Pred(x)	returns the predecessor of x
Pred(x, n)	returns the n'th predecessor of x

The operations "inc," "dec," "succ," and "pred" in ordinal types may fail and raise a "RangeDefect" or "OverflowDefect" exception. These exceptions occur when the code has been compiled with the appropriate runtime checks enabled.

Subranges

A subrange type represents a range of values from an integer or enumeration type (the base type). Example:

Nim Codes

```nim
type
  MySubrange = range[0..5]
```

"MySubrange" defines a subrange of the "int" type that can only store values from 0 to 5. Assigning any other value to a variable of type "MySubrange" will result in a compile-time or runtime error. However, assignments between the base type and its subrange types are allowed.

In the system module, there is a fundamental type called "Natural," which is defined as "range[0..high(int)]". The "high" function returns the maximum value of the "int" type. While other programming languages may recommend using unsigned integers for natural numbers, this approach is often not advisable. Unsigned arithmetic, which wraps around, is not desirable simply because the numbers cannot be negative.

Nim's "Natural" type helps prevent this common programming mistake.

Sets

In Nim, the set type represents the mathematical concept of a set. It requires the base type to be an ordinal type of a specific size. Supported base types include:

- Signed integers: int8 to int16
- Unsigned integers: uint8/byte to uint16
- Characters: char
- Enumerations
- Ordinal subrange types, such as range[-10..10]

When constructing a set using signed integer literals, the base type of the set is defined within the range of 0 to DefaultSetElements-1. The value of DefaultSetElements is always set to 2^8. The maximum range length for the base type of a set is MaxSetElements, which is consistently 2^{16}. If a type with a larger range length is used, it will be coerced into the range of 0 to MaxSetElements-1.

These limitations exist because sets in Nim are implemented as high-performance bit vectors. Declaring a set with a larger type than the supported range will raise an error.

Nim Codes

```
var s: set[int64] # Error: set is too large; use `std/sets` for ordinal types
                  # with more than 2^16 elements
```

Note: In addition to the standard sets in Nim, there are also hash sets available (imported using import std/sets). Hash sets do not have the same restrictions as regular sets.

Sets in Nim can be constructed using the set constructor. The empty set is represented by {}. The set constructor can be used to include specific elements or ranges of elements:

Nim Codes

```
type
  CharSet = set[char]
var
```

x: CharSet

x = {'a'..'z', '0'..'9'} # This constructs a set that contains the

                 # letters from 'a' to 'z' and the digits

                 # from '0' to '9'

The module std/setutils provides a way to initialize a set from an iterable:

Nim Codes

```
import std/setutils

let uniqueChars = myString.toSet
```

These operations are supported by sets:

Operation	meaning
A + B	union of two sets
A * B	intersection of two sets
A - B	difference of two sets (A without B's elements)
A == B	set equality
A <= B	subset relation (A is subset of B or

	equal to B)
A < B	strict subset relation (A is a proper subset of B)
e in A	set membership (A contains element e)
e notin A	A does not contain element e
contains (A, e)	A contains element e
card (A)	the cardinality of A (number of elements in A)
incl (A, elem)	same as A = A + {elem}
excl(A, elem)	same as A = A - {elem}

Bit fields

Sets are commonly used to define procedure flags, offering a cleaner and safer alternative to combining integer constants with bitwise OR operations.

Enums, sets, and casting can be effectively combined, as shown below:

Nim Codes

```
type
  MyFlag* {.size: sizeof(cint).} = enum
    A
    B
    C
    D
  MyFlags = set[MyFlag]

proc toNum(f: MyFlags): int = cast[cint](f)
proc toFlags(v: int): MyFlags = cast[MyFlags](v)

assert toNum({}) == 0
assert toNum({A}) == 1
assert toNum({D}) == 8
assert toNum({A, C}) == 5
assert toFlags(0) == {}
assert toFlags(7) == {A, B, C}
```

It's important to note that enum values are transformed into powers of 2 when used with sets.

When working with enums, sets, and C interoperability, it is recommended to utilize distinct cint types.

To enhance compatibility with C, consider using the bitsize pragma.

Arrays

Arrays are straightforward containers with a fixed length, where all elements share the same type. The index type can be any ordinal type.

You can create arrays using the [] notation:

```nim
Nim Codes
type
  IntArray = array[0..5, int] # an array that is indexed with 0..5
var
  x: IntArray
x = [1, 2, 3, 4, 5, 6]
for i in low(x) .. high(x):
  echo x[i]
```

Accessing the i-th element of an array x is done using the notation x[i]. Array access is always checked for bounds, either at compile-time or runtime. If desired, these checks can be disabled using pragmas or the --bound_checks:off command line switch.

Arrays in Nim are value types, meaning the entire contents are copied when using the assignment operator.

The len procedure returns the length of an array. Additionally, low(a) returns the lowest valid index, while high(a) returns the highest valid index.

Nim Codes

```
type
  Direction = enum
    north, east, south, west
  BlinkLights = enum
    off, on, slowBlink, mediumBlink, fastBlink
  LevelSetting = array[north..west, BlinkLights]
var
  level: LevelSetting
level[north] = on
level[south] = slowBlink
level[east] = fastBlink
echo level        # --> [on, fastBlink, slowBlink,
```

off]
echo low(level)   # --> north
echo len(level)   # --> 4
echo high(level) # --> west

In Nim, nested arrays (multidimensional arrays) can have different dimensions with varying index types. This flexibility allows for a slightly different nesting syntax.

Expanding on the previous example, where a level was defined as an array of enums indexed by another enum, we can introduce additional lines of code to incorporate a light tower type that is subdivided into height levels accessible through integer indices:

Nim Codes

```
type
  LightTower = array[1..10, LevelSetting]
var
  tower: LightTower
tower[1][north] = slowBlink
tower[1][east] = mediumBlink
```

```
echo len(tower)      # --> 10
echo len(tower[1])   # --> 4
echo  tower                    # --> [[slowBlink, mediumBlink, ...more output..
# The following lines don't compile due to type mismatch errors
#tower[north][east] = on
#tower[0][1] = on
```

It's important to note that the len procedure in Nim only returns the length of the array's first dimension. To better illustrate the nested nature of the LightTower, an alternative way to define it is by embedding the LevelSetting type directly as the type of the first dimension:

Nim Codes

```
type
  LightTower = array[1..10, array[north..west, BlinkLights]]
```

In many programming languages, arrays conventionally start at zero. In Nim, you can use a syntax shortcut to specify a range that starts from zero and ends at the specified index minus one:

Nim Codes

```
type
  IntArray = array[0..5, int]   # an array that is indexed with 0..5
  QuickArray = array[6, int]    # an array that is indexed with 0..5
var
  x: IntArray
  y: QuickArray
x = [1, 2, 3, 4, 5, 6]
y = x
for i in low(x) .. high(x):
  echo x[i], y[i]
```

Sequences

Sequences share similarities with arrays but have the advantage of dynamic length, allowing them to be resized during runtime, similar to strings. Sequences are always allocated on the heap and managed by garbage collection.

Sequences are indexed starting from position 0 using an int. They support len, low, and high operations, similar to arrays. To access the i-th element of a sequence x, you can use the notation x[i].

You can create sequences using the array constructor [] combined with the array-to-sequence operator @. Alternatively, you can allocate space for a sequence using the newSeq procedure.

Sequences can be passed to openarray parameters.

Example:

Nim Codes

```
var
  x: seq[int]  # a reference to a sequence of integers
x = @[1, 2, 3, 4, 5, 6] # the @ turns the array into a sequence allocated on the heap
```

To initialize sequence variables in Nim, use the notation @[].

The for statement in Nim can be used with one or two variables when working with sequences. In the one-variable form, the variable holds the value provided by the sequence. This form of the for statement loops over the results obtained from the items() iterator in the system module. In the two-variable form, the index position is assigned to the first variable, and the corresponding value is assigned to the second variable. In this case, the for statement loops over the results obtained from the pairs() iterator in the system module. Here are some examples:

Nim Codes

```
for value in @[3, 4, 5]:
  echo value
# --> 3
# --> 4
# --> 5
```

```
for i, value in @[3, 4, 5]:
  echo "index: ", $i, ", value:", $value
# --> index: 0, value:3
# --> index: 1, value:4
# --> index: 2, value:5
```

Open arrays

Note: Open arrays are only applicable for parameters.

Fixed-size arrays can be too rigid in situations where procedures need to handle arrays of varying sizes. This is where the open array type comes into play. Open arrays are always indexed starting from position 0 using an int. Operations such as len, low, and high are available for open arrays. The beauty of open arrays is that they can accept any array with a compatible base type as a parameter, regardless of the index type used.

Nim Codes

```
var
  fruits: seq[string]       # reference to a sequence of strings that is initialized with '@[]'
```

```
capitals: array[3, string]  # array of strings with a fixed size

capitals = ["New York", "London", "Berlin"]   # array 'capitals' allows assignment of only three elements
fruits.add("Banana")         # sequence 'fruits' is dynamically expandable during runtime
fruits.add("Mango")

proc openArraySize(oa: openArray[string]): int =
  oa.len

assert openArraySize(fruits) == 2    # procedure accepts a sequence as parameter
assert openArraySize(capitals) == 3   # but also an array type
```

Nested open arrays, also known as multidimensional open arrays, are not supported in Nim. This limitation arises from the fact that the requirement for nested open arrays is rare, and implementing them efficiently poses significant challenges. Consequently, the Nim programming language does not offer native support for multidimensional open arrays.

Objects

The object type in Nim serves as the default choice for encapsulating multiple values within a named structure. Objects are considered value types, meaning that when an object is assigned to a new variable, all of its components are copied as well.

For each object type 'Foo', there exists a constructor 'Foo(field: value, ...)' that allows the initialization of its fields. Any unspecified fields will be assigned their respective default values.

Varargs

Nim has a concept called varargs parameters, which operate similarly to openarray parameters. Varargs parameters allow passing a variable number of arguments to a procedure. The compiler automatically converts the list of arguments into an array.

Nim Codes

```
proc myWriteln(f: File, a: varargs[string]) =
  for s in items(a):
    write(f, s)
  write(f, "\n")

myWriteln(stdout, "abc", "def", "xyz")
# is transformed by the compiler to:
myWriteln(stdout, ["abc", "def", "xyz"])
```

In Nim, the automatic transformation of varargs parameters into arrays is only applied when the varargs parameter is the last parameter in the procedure header. It's important to note that type conversions can also be performed within this context.

Nim Codes

```
proc myWriteln(f: File, a: varargs[string, `$`]) =
  for s in items(a):
    write(f, s)
```

write(f, "\n")

myWriteln(stdout, 123, "abc", 4.0)
# is transformed by the compiler to:
myWriteln(stdout, [$123, $"abc", $4.0])

In the given example, the $ symbol is applied to each argument passed to the parameter 'a'. However, when applied to strings, the $ operator has no effect (it is essentially a no-op).

Slices

Slices in Nim have a syntax similar to subrange types but serve a different purpose. A slice is an object of the Slice type, consisting of two bounds, 'a' and 'b'. On its own, a slice may not be particularly useful. However, other collection types define operators that accept Slice objects to define ranges.

Nim Codes

```
var
  a = "Nim is a programming language"
```

```
b = "Slices are useless."

echo a[7 .. 12] # --> 'a prog'
b[11 .. ^2] = "useful"
echo b # --> 'Slices are useful.'
```

In the previous example, slices are used to modify a specific portion of a string. The bounds of a slice can accommodate any value supported by their respective type. However, it is the procedure utilizing the slice object that determines which values are accepted.

To understand the various ways of specifying indices for strings, arrays, sequences, and similar structures in Nim, it's important to remember that Nim follows a zero-based indexing convention.

Nim Codes

```
"Slices are useless."
 |       |   |
 0       11  17   using indices
 ^19     ^8  ^2   using ^ syntax
```

In the provided example, the expression 'b[0 .. ^1]' is equivalent to 'b[0 .. b.len-1]' and 'b[0 ..< b.len]'. The notation '^1' is a shorthand way to specify 'b.len-1' using the backwards index operator.

In the given scenario, assuming the string 'b' ends with a period, we can extract the portion of the string that is considered "useless" by using 'b[11 .. ^2]'. By assigning "useful" to 'b[11 .. ^2]', we replace the "useless" portion with "useful," resulting in the string "Slices are useful."

It's worth noting that alternate ways to express this operation include 'b[^8 .. ^2] = "useful"', 'b[11 .. b.len-2] = "useful"', or 'b[11 ..< b.len-1] = "useful"'. Additionally, by defining a constant 'lastIndex' as 'const lastIndex = ^1', we can later use it as 'b[0 .. lastIndex]' to refer to the last index of the string.

Objects

In Nim, the object type is the default choice for encapsulating multiple values within a named structure. Objects are considered value types, which means that when an object is assigned to a new variable, all of its components are copied as well.

For each object type 'Foo', there exists a constructor 'Foo(field: value, ...)' that allows initialization of its fields. Any unspecified fields will be assigned their respective default values.

Nim Codes

```nim
type
  Person = object
    name: string
    age: int

var person1 = Person(name: "Peter", age: 30)

echo person1.name # "Peter"
echo person1.age  # 30
```

```
var person2 = person1 # copy of person 1

person2.age += 14

echo person1.age # 30
echo person2.age # 44


# the order may be changed
let person3 = Person(age: 12, name: "Quentin")

# not every member needs to be specified
let person4 = Person(age: 3)
# unspecified members will be initialized with
their default
# values. In this case it is the empty string.
doAssert person4.name == ""
```

To make object fields visible from outside the module where they are defined, they need to be marked with an asterisk (*). This notation indicates that the fields are accessible and can be referenced externally.

> **Nim Codes**
>
> type
>   Person* = object # the type is visible from other modules
>   name*: string # the field of this type is visible from other modules
>   age*: int

Tuples

Tuples in Nim behave similarly to objects in terms of their behavior. They are value types, so when the assignment operator is used, each component of the tuple is copied. However, there are some differences. Tuple types in Nim are structurally typed, meaning that different tuple types are considered equivalent if they specify fields of the same type and with the same name in the same order.

To create tuples, the constructor '()' can be used. It's important to ensure that the order of the fields in the constructor matches the order specified in the tuple's definition. Unlike objects, a specific name for the tuple type is not used in this context.

Similar to object types, the notation 't.field' is used to access a field within a tuple. However, unlike objects, tuples provide an additional notation 't[i]' to access the i-th field. It's important to note that 'i' must be a constant integer when using this notation.

Nim Codes

```
type
  # type representing a person:
  # A person consists of a name and an age.
  Person = tuple
    name: string
    age: int

  # Alternative syntax for an equivalent type.
```

```nim
PersonX = tuple[name: string, age: int]

# anonymous field syntax
PersonY = (string, int)

var
  person: Person
  personX: PersonX
  personY: PersonY

person = (name: "Peter", age: 30)
# Person and PersonX are equivalent
personX = person

# Create a tuple with anonymous fields:
personY = ("Peter", 30)

# A tuple with anonymous fields is compatible with a tuple that has
# field names.
person = personY
personY = person

# Usually used for short tuple initialization
```

```
syntax
person = ("Peter", 30)

echo person.name # "Peter"
echo person.age  # 30

echo person[0] # "Peter"
echo person[1] # 30

# You don't need to declare tuples in a separate type section.
var building: tuple[street: string, number: int]
building = ("Rue del Percebe", 13)
echo building.street

# The following line does not compile, they are different tuples!
#person = building
# --> Error: type mismatch: got (tuple[street: string, number: int])
#     but expected 'Person'
```

Tuples can be unpacked during variable assignment, allowing individual fields of the tuple to be assigned to named variables. This can be convenient when direct assignment of tuple fields is desired. To ensure correct behavior, parentheses must be used around the values intended for unpacking. Otherwise, the same value will be assigned to all individual variables.

Nim Codes

```
type
  Node = ref object
    le, ri: Node
    data: int

var n = Node(data: 9)
echo n.data
# no need to write n[].data; in fact n[].data is highly discouraged!
```

To allocate a new traced object, you can utilize the built-in procedure new. The new procedure allows you to create and allocate memory for a new object of a specified type. This is particularly useful for creating traced objects that will be managed by the garbage collector. By invoking new, you can dynamically allocate memory for the object and initialize it appropriately.

Nim Codes

var n: Node
new(n)

To effectively handle untraced memory, you can utilize the alloc, dealloc, and realloc procedures. For more detailed information on these procedures, you can refer to the documentation provided in the system module.

In Nim, the value of a reference that doesn't point to a specific object or memory location is represented by nil. This convenient representation enables the identification of uninitialized references or references that don't point to valid data.

Procedural type in NIM

In Nim, a procedural type refers to a pointer to a procedure, although it possesses an abstract nature. It's worth noting that nil is a valid value for a procedural type variable. Nim leverages procedural types to incorporate functional programming techniques.

Here's an example:

Nim Codes

```nim
proc greet(name: string): string =
  "Hello, " & name & "!"

proc bye(name: string): string =
  "Goodbye, " & name & "."

proc communicate(greeting: proc (x: string): string, name: string) =
  echo greeting(name)

communicate(greet, "John")
communicate(bye, "Mary")
```

When working with procedural types, it's crucial to consider that the calling convention of a procedure affects the compatibility of procedural types. Two procedural types are deemed compatible only if they share the same calling convention. The manual provides an outline of the various calling conventions available.

Distinct Type

A distinct type allows for the creation of a new type that doesn't establish a subtype relationship with its base type. It necessitates explicitly defining all the behavior for the distinct type. To facilitate this, both the distinct type and its base type can be cast to one another. The manual offers examples to illustrate this concept.

Modules

In Nim, it's possible to divide a program into separate pieces using the module concept, with each module residing in its own file. Modules provide the benefits of information hiding and separate compilation. By utilizing the import statement, a module can access the symbols defined in another module. It's important to note that only top-level symbols marked with an asterisk (*) are exported and made accessible outside of the module.

Nim Codes

```nim
# Module A
var
  x*, y: int

proc `*` *(a, b: seq[int]): seq[int] =
  # allocate a new sequence:
  newSeq(result, len(a))
  # multiply two int sequences:
  for i in 0 ..< len(a): result[i] = a[i] * b[i]

when isMainModule:
  # test the new `*` operator for sequences:
  assert(@[1, 2, 3] * @[1, 2, 3] == @[1, 4, 9])
```

The above module exports x and all top-level symbols denoted by *, but y is not exported.

When a program starts, a module's top-level statements are executed. This feature can be utilized to initialize complex data structures, among other things.

Every module contains a special magic constant called isMainModule, which evaluates to true if the module is compiled as the main file. This constant proves useful for embedding tests within the module, as demonstrated in the previous example.

To qualify a symbol with its corresponding module, the syntax module.symbol can be employed. If a symbol is ambiguous, meaning it is defined in two or more different modules that are both imported by a third module, it must be qualified to disambiguate its usage.

Nim Codes
```
# Module A
var x*: string
```

Nim Codes
```
# Module B
var x*: int
```

Nim Codes

```nim
# Module C
import A, B
write(stdout, x) # error: x is ambiguous
write(stdout, A.x) # okay: qualifier used

var x = 4
write(stdout, x) # not ambiguous: uses the module C's x
```

However, this rule does not apply to procedures or iterators. Here, the overloading rules come into play:

Nim Codes
```nim
# Module A
proc x*(a: int): string = $a
```

Nim Codes
```nim
# Module B
proc x*(a: string): string = $a
```

Nim Codes
```nim
# Module C
```

```
import A, B
write(stdout, x(3))   # no error: A.x is called
write(stdout, x(""))  # no error: B.x is called

proc x*(a: int): string = discard
write(stdout, x(3))   # ambiguous: which `x` is to call?
```

Excluding Symbols

The standard import statement allows for the inclusion of all exported symbols from a module. However, it's possible to limit the imported symbols by explicitly specifying the symbols to be excluded using the except qualifier.

Nim Codes
```
import mymodule except y
```

Form Statement

Previously, we encountered the basic import statement that imports all exported symbols from a module. An alternative to this is the from import statement, which allows for the selective importation of specific listed symbols.

> **Nim Codes**
> from mymodule import x, y, z

The from statement also has the ability to enforce namespace qualification on symbols. This means that symbols can be made available through the from import, but they must be explicitly qualified to be used.

> **Nim Codes**
> from mymodule import x, y, z
>
> x()   # use x without any qualification

> **Nim Codes**
> from mymodule import nil
>
> mymodule.x() # must qualify x with the module name as prefix
>
> x()   # using x here without qualification is

a compile error

To address the issue of long module names, it's possible to define a shorter alias that can be used to qualify symbols instead. This provides a more concise way of referencing symbols from a module without repeatedly using the lengthy module name for qualification.

Nim Codes

```nim
from mymodule as m import nil

m.x()      # m is aliasing mymodule
```

Include Statement

The purpose of the include statement differs fundamentally from importing a module. Rather than importing the module itself, it directly incorporates the contents of a file. The include statement is especially valuable for breaking down a large module into multiple files, promoting improved code organization and modularity.

Nim Codes

include fileA, fileB, fileC

PART TWO

This document acts as a tutorial for the advanced constructs of the Nim programming language. It's crucial to acknowledge that the information contained herein might be slightly outdated, as the manual offers a more comprehensive array of examples that demonstrate the advanced language features.

Programs

Pragmas in Nim serve as a convenient way to provide the compiler with additional information or commands without the need for introducing numerous new keywords. Enclosed within special curly dot brackets, {.} and .{}, pragmas offer a concise syntax. However, it's important to note that this tutorial does not delve into the topic of pragmas. For a comprehensive understanding of available pragmas, it is recommended to refer to the manual or user guide.

Object-Oriented Programming

Nim offers minimalistic support for object-oriented programming (OOP) while still allowing for the utilization of powerful OOP techniques. While OOP is considered one approach to program design, it's essential to recognize that it is not the only approach. In many cases, a procedural approach can lead to simpler and more efficient code. Specifically, favoring composition over inheritance is often a better design choice.

Inheritance

In Nim, inheritance is entirely optional. To enable inheritance with runtime type information, an object needs to inherit from RootObj. This can be achieved directly or indirectly by inheriting from an object that itself inherits from RootObj. Typically, types that involve inheritance are also marked as reference (ref) types, even though it is not strictly enforced. To check if an object is of a specific type at runtime, the of operator can be used.

Nim Codes

```
type
  Person = ref object of RootObj
    name*: string   # the * means that `name` is accessible from other modules
    age: int        # no * means that the field is hidden from other modules
```

```
  Student = ref object of Person  # Student inherits from Person
    id: int                       # with an id field

var
  student: Student
  person: Person
assert(student of Student) # is true
# object construction:
student = Student(name: "Anton", age: 5, id: 2)
echo student[]
```

In Nim, inheritance is achieved using the "object of" syntax. Currently, multiple inheritance is not supported. If an object type lacks a suitable ancestor, the convention is to use RootObj as its ancestor, although this is not mandatory. Objects without ancestors are implicitly considered final. To introduce new object roots other than system.RootObj, the "inheritable" pragma can be utilized. This can be seen in the GTK wrapper, for instance.

When utilizing inheritance, it is recommended to use ref objects. While not strictly necessary, with non-ref objects, assignments such as "let person: Person = Student(id: 123)" will truncate subclass fields.

Note: For simple code reuse, composition (has-a relation) is often preferable to inheritance (is-a relation). Since objects are value types in Nim, composition is just as efficient as inheritance.

Mutually recursive types

Objects, tuples, and references have the capability to represent complex data structures that depend on each other. These types are mutually recursive. In Nim, it is only possible to declare such types within a single type section. Declaring them elsewhere would require arbitrary symbol lookahead, which would negatively impact compilation speed.

Example:

```
type
  Node = ref object   # a reference to an object with the following field:
    le, ri: Node    # left and right subtrees
    sym: ref Sym    # leaves contain a reference to a Sym

  Sym = object      # a symbol
    name: string    # the symbol's name
    line: int       # the line the symbol was declared in
    code: Node      # the symbol's abstract syntax tree
```

Type Conversions

In Nim, a clear distinction exists between type casts and type conversions. Type casts are accomplished using the cast operator, which instructs the compiler to interpret a bit pattern as another type.

On the other hand, type conversions offer a more considerate approach to converting one type to another. They preserve the abstract value rather than the bit pattern. If a type conversion is not possible, the compiler will raise an error or exception.

The syntax for type conversions is destination_type(expression_to_convert), resembling a regular function call:

Nim Codes
```
proc getID(x: Person): int =
  Student(x).id
```

If x is not a Student, the InvalidObjectConversionDefect exception is raised.

Object Variants

In certain situations, an object hierarchy can be excessive, and a simpler variant type is more appropriate.

For example:

Nim Codes
```
# This is an example how an abstract syntax tree
could be modelled in Nim
```

```
type
  NodeKind = enum  # the different node types
    nkInt,       # a leaf with an integer value
    nkFloat,     # a leaf with a float value
    nkString,    # a leaf with a string value
    nkAdd,       # an addition
    nkSub,       # a subtraction
    nkIf         # an if statement
  Node = ref object
    case kind: NodeKind  # the `kind` field is the discriminator
    of nkInt: intVal: int
    of nkFloat: floatVal: float
    of nkString: strVal: string
    of nkAdd, nkSub:
      leftOp, rightOp: Node
    of nkIf:
      condition, thenPart, elsePart: Node

var n = Node(kind: nkFloat, floatVal: 1.0)
# the following statement raises an `FieldDefect` exception, because
# n.kind's value does not fit:
n.strVal = ""
```

As demonstrated in the example, one advantage of an object hierarchy is that no conversion between different object types is required. However, accessing invalid object fields will result in an exception being raised.

Method Call Syntax

Nim provides syntactic sugar for calling routines. The syntax obj.methodName(args) can be used instead of methodName(obj, args). If there are no additional arguments, the parentheses can be omitted: obj.len instead of len(obj).

This method call syntax is not limited to objects; it can be used with any type:

Nim Codes

```
import std/strutils

echo "abc".len # is the same as echo len("abc")
echo "abc".toUpperAscii()
echo({'a', 'b', 'c'}.card)
```

stdout.writeLine("Hallo")   # the same as writeLine(stdout, "Hallo")

(Another perspective on the method call syntax is that it offers the missing postfix notation.)

Thus, writing "pure object-oriented" code is straightforward:

Nim Codes

import std/[strutils, sequtils]

stdout.writeLine("Give a list of numbers (separated by spaces): ")
stdout.write(stdin.readLine.splitWhitespace.map(parseInt).max.`$`)
stdout.writeLine(" is the maximum!")

Properties

As illustrated in the example above, Nim does not require the use of get-properties. The same effect can be achieved by using ordinary get-procedures called with the method call syntax. However, when it comes to setting a value, a special setter syntax is required:

Nim Codes

```
type
  Socket* = ref object of RootObj
    h: int # cannot be accessed from the outside of the module due to missing star

proc `host=`*(s: var Socket, value: int) {.inline.} =
  ## setter of host address
  s.h = value

proc host*(s: Socket): int {.inline.} =
  ## getter of host address
  s.h

var s: Socket
new s
s.host = 34  # same as `host=`(s, 34)
```

(The example also shows inline procedures.)

The [] array access operator can be overloaded to provide array properties:

Nim Codes
```nim
type
  Vector* = object
    x, y, z: float

proc `[]=`* (v: var Vector, i: int, value: float) =
  # setter
  case i
  of 0: v.x = value
  of 1: v.y = value
  of 2: v.z = value
  else: assert(false)

proc `[]`* (v: Vector, i: int): float =
  # getter
  case i
  of 0: result = v.x
  of 1: result = v.y
  of 2: result = v.z
  else: assert(false)
```

The example provided is not particularly useful since a vector is better represented by a tuple, which already provides access through v[].

Dynamic Dispatch

Procedures in Nim always utilize static dispatch. To achieve dynamic dispatch, the "proc" keyword can be replaced with "method":

Nim Codes

```nim
type
  Expression = ref object of RootObj ## abstract base class for an expression
  Literal = ref object of Expression
    x: int
  PlusExpr = ref object of Expression
    a, b: Expression

# watch out: 'eval' relies on dynamic binding
method eval(e: Expression): int {.base.} =
  # override this base method
  quit "to override!"
```

```
method eval(e: Literal): int = e.x
method eval(e: PlusExpr): int = eval(e.a) + eval(e.b)

proc newLit(x: int): Literal = Literal(x: x)
proc newPlus(a, b: Expression): PlusExpr = PlusExpr(a: a, b: b)

echo eval(newPlus(newPlus(newLit(1), newLit(2)), newLit(4)))
```

Please note that in the provided example, the constructors newLit and newPlus are declared as procs. This choice is based on the rationale that static binding is more appropriate for them. However, eval is defined as a method since it requires dynamic binding.

Note: Starting from Nim 0.20, the --multimethods:on flag must be explicitly passed during compilation to enable the use of multi-methods.

In a multi-method, all parameters with an object type are used for dispatching purposes:

```nim
type
  Thing = ref object of RootObj
  Unit = ref object of Thing
    x: int

method collide(a, b: Thing) {.inline.} =
  quit "to override!"

method collide(a: Thing, b: Unit) {.inline.} =
  echo "1"

method collide(a: Unit, b: Thing) {.inline.} =
  echo "2"

var a, b: Unit
new a
new b
collide(a, b) # output: 2
```

As shown in the example, the invocation of a multi-method must not be ambiguous. In the case of Collide, the implementation of Collide2 takes precedence over Collide1 due to the resolution process working from left to right. Consequently, the combination of Unit, Thing is favored over Thing, Unit.

Performance note: Nim optimizes method calls by generating dispatch trees instead of virtual method tables. This approach eliminates the costly indirect branch associated with method calls and allows for inlining. However, it is important to note that certain optimizations, such as compile-time evaluation or dead code elimination, do not apply to methods.

Exceptions

In Nim, exceptions are represented as objects. It is customary to suffix exception types with 'Error'. The system module defines an exception hierarchy that is recommended to be followed. Exceptions inherit from system.Exception, which provides a common interface.

Since the lifetime of exceptions is uncertain, they need to be allocated on the heap. The compiler prohibits raising an exception created on the stack. When raising exceptions, it is important to include a descriptive reason in the msg field.

It is customary to raise exceptions only in exceptional circumstances and not to use them as an alternative control flow method.

Raise Statement

The act of raising an exception is accomplished using the raise statement:

Nim Codes
```
var
  e: ref OSError
new(e)
e.msg = "the request to the OS failed"
raise e
```

When the raise keyword is not followed by an expression, it signifies the re-raising of the last exception. To prevent the repetition of this common code pattern, the newException template in the system module can be utilized. Its purpose is to facilitate the creation of exceptions.

Nim Codes

raise newException(OSError, "the request to the OS failed")

Try Statement

The try statement in the Nim language handles exceptions:

Nim Codes

```
from std/strutils import parseInt

# read the first two lines of a text file that should contain numbers
# and tries to add them
var
  f: File
if open(f, "numbers.txt"):
  try:
    let a = readLine(f)
    let b = readLine(f)
```

```
    echo "sum: ", parseInt(a) + parseInt(b)
  except OverflowDefect:
    echo "overflow!"
  except ValueError:
    echo "could not convert string to integer"
  except IOError:
    echo "IO error!"
  except CatchableError:
    echo "Unknown exception!"
    # reraise the unknown exception:
    raise
  finally:
    close(f)
```

The statements following the try block are executed unless an exception is raised. In such cases, the corresponding except block is executed.

The empty except block is triggered when there is an exception that is not explicitly listed. It is analogous to an else block in if statements.

If a finally block is present, it is always executed after the exception handlers.

An exception is handled within an except block. If an exception is left unhandled, it propagates through the call stack. This often means that the remaining portion of the procedure, not within a finally block, is not executed when an exception occurs.

To access the actual exception object or message within an except block, you can utilize the getCurrentException() and getCurrentExceptionMsg() procs from the system module. Here's an example:

Nim Codes

```nim
try:
  doSomethingHere()
except CatchableError:
  let
    e = getCurrentException()
    msg = getCurrentExceptionMsg()
  echo "Got exception ", repr(e), " with message ", msg
```

Annotating procs with raised exceptions

By using the optional {.raises.} pragma, you can indicate the intended exceptions raised by a proc or specify that it raises no exceptions at all. When utilizing the {.raises.} pragma, the compiler will verify its accuracy. For instance, if you declare that a proc raises IOError, but at some point it (or one of the procs it calls) starts raising a different exception, the compiler will halt the compilation of that proc. Here's an example of its usage:

```nim
proc complexProc() {.raises: [IOError, ArithmeticDefect].} =
    ...

proc simpleProc() {.raises: [].} =
    ...
```

Once you have implemented code with the mentioned structure, any modifications to the list of raised exceptions will trigger a compiler error. This error message will specify the line of the proc that failed to validate the pragma, along with the uncaught exception being raised, along with the file and line where it occurs. This information can help identify the problematic code that has been altered.

If you want to add the {.raises.} pragma to existing code, the compiler can assist you in this regard. By including the {.effects.} pragma statement in your proc, the compiler will output all inferred effects up to that point. It's worth noting that exception tracking is part of Nim's effect system. Alternatively, you can use the Nim doc command to generate documentation for an entire module, which will annotate all procs with the list of raised exceptions. For further information on Nim's effect system and related pragmas, please refer to the manual.

Generics

Generics provide a means of parameterizing procs, iterators, or types with type parameters in Nim. These parameters are enclosed within square brackets, such as Foo[T]. Generics prove to be highly useful when creating efficient and type-safe containers:

Nim Codes

```nim
type
  BinaryTree*[T] = ref object # BinaryTree is a generic type with
                # generic param `T`
    le, ri: BinaryTree[T]    # left and right subtrees; may be nil
    data: T                  # the data stored in a node

proc newNode*[T](data: T): BinaryTree[T] =
  # constructor for a node
  new(result)
  result.data = data

proc add*[T](root: var BinaryTree[T], n: BinaryTree[T]) =
  # insert a node into the tree
  if root == nil:
    root = n
  else:
```

```
  var it = root
  while it != nil:
    # compare the data items; uses the generic `cmp` proc
    # that works for any type that has a `==` and `<` operator
    var c = cmp(it.data, n.data)
    if c < 0:
      if it.le == nil:
        it.le = n
        return
      it = it.le
    else:
      if it.ri == nil:
        it.ri = n
        return
      it = it.ri

proc add*[T](root: var BinaryTree[T], data: T) =
  # convenience proc:
  add(root, newNode(data))

iterator preorder*[T](root: BinaryTree[T]): T =
  # Preorder traversal of a binary tree.
```

```
  # This uses an explicit stack (which is more efficient than
  # a recursive iterator factory).
  var stack: seq[BinaryTree[T]] = @[root]
  while stack.len > 0:
    var n = stack.pop()
    while n != nil:
      yield n.data
      add(stack, n.ri)  # push right subtree onto the stack
      n = n.le          # and follow the left pointer

var
  root: BinaryTree[string]  # instantiate a BinaryTree with `string`
add(root, newNode("hello"))  # instantiates `newNode` and `add`
add(root, "world")           # instantiates the second `add` proc
for str in preorder(root):
  stdout.writeLine(str)
```

116

The provided example demonstrates a generic binary tree. The brackets are used differently depending on the context, either to introduce type parameters or to instantiate a generic proc, iterator, or type. As shown in the example, generics support overloading, allowing the best match for the add proc to be used. The built-in add procedure for sequences is not hidden and is employed in the preorder iterator.

When using generics with the method call syntax, there is a special [:T] syntax:

Nim Codes

```nim
proc foo[T](i: T) =
  discard

var i: int

# i.foo[int]() # Error: expression 'foo(i)' has no type (or is ambiguous)

i.foo[:int]() # Success
```

Templates

Templates in Nim are a straightforward substitution mechanism that operates on the abstract syntax trees of the language. They are processed during the semantic pass of the compiler. Templates seamlessly integrate with the rest of the language and do not suffer from the shortcomings of C's preprocessor macros.

To invoke a template, simply call it like a procedure.

Here's an example:

Nim Codes
```nim
template `!=` (a, b: untyped): untyped =
  # this definition exists in the System module
  not (a == b)


assert(5 != 6) # the compiler rewrites that to:
assert(not (5 == 6))
```

The templates used for the !=, >, >=, in, notin, and isnot operators offer the advantage that if you overload the == operator, the != operator is automatically available and behaves correctly. However, it's important to note that this behavior does not apply to IEEE floating-point numbers, as NaN values can disrupt basic boolean logic.

The expression a > b is transformed into b < a. Similarly, a in b is transformed into contains(b, a). The operators!=, >, >=, in, notin, and isnot have their intuitive meanings.

Templates are particularly valuable for implementing lazy evaluation. Let's consider a simple logging procedure as an example:

Nim Codes

```nim
const
  debug = true

proc log(msg: string) {.inline.} =
  if debug: stdout.writeLine(msg)

var
  x = 4
log("x has the value: " & $x)
```

The provided code has a limitation: even if the debug flag is set to false, the expensive $ and & operations are still executed. This is because procedure argument evaluation occurs eagerly.

However, this issue can be addressed by transforming the log procedure into a template, which solves the problem:

Nim Codes

```
const
  debug = true

template log(msg: string) =
  if debug: stdout.writeLine(msg)

var
  x = 4
log("x has the value: " & $x)
```

The types of the parameters in templates can be regular types or the meta types untyped, typed, or type. Using the type meta type indicates that only a type symbol is allowed as an argument, while untyped means that symbol lookups and type

120

resolution are not performed before the expression is passed to the template.

If a template does not have an explicitly specified return type, the type void is used to maintain consistency with procs and methods.

To pass a block of statements to a template, the untyped keyword should be used for the last parameter:

Nim Codes

```
template withFile(f: untyped, filename: string, mode: FileMode,
          body: untyped) =
  let fn = filename
  var f: File
  if open(f, fn, mode):
    try:
      body
    finally:
      close(f)
  else:
    quit("cannot open: " & fn)
```

```
withFile(txt, "ttempl3.txt", fmWrite):
  txt.writeLine("line 1")
  txt.writeLine("line 2")
```

In the provided example, the two writeLine statements are associated with the body parameter. The withFile template includes boilerplate code that helps avoid a common error: forgetting to close the file. Notice how the statement let fn = filename ensures that filename is evaluated only once.

Example: Lifting Procs

Nim Codes
```
import std/math

template liftScalarProc(fname) =
  ## Lift a proc taking one scalar parameter and returning a
  ## scalar value (eg `proc sssss[T](x: T): float`),
  ## to provide templated procs that can handle a single
```

```nim
## parameter of seq[T] or nested seq[seq[]] or the same type
##
## ```Nim
## liftScalarProc(abs)
##   # now abs(@[@[1,-2], @[-2,-3]]) == @[@[1,2], @[2,3]]
## ```
proc fname[T](x: openarray[T]): auto =
  var temp: T
  type outType = typeof(fname(temp))
  result = newSeq[outType](x.len)
  for i in 0..<x.len:
    result[i] = fname(x[i])

liftScalarProc(sqrt)   # make sqrt() work for sequences
echo sqrt(@[4.0, 16.0, 25.0, 36.0])   # => @[2.0, 4.0, 5.0, 6.0]
```

Compilation to JavaScript

When converting Nim code to JavaScript, it's important to consider the following points:

- The addr and ptr constructs have slightly different semantic meanings in JavaScript. If you're not familiar with their translation to JavaScript, it's advisable to avoid using them.
- The castT operation in JavaScript is translated as (x), except when casting between signed and unsigned integers, where it behaves similarly to a static cast in the C language.
- In JavaScript, the term cstring refers to a JavaScript string. It's recommended to use cstring only when it makes semantic sense. For instance, it's best to avoid using cstring as a binary data buffer.

PART THREE

This tutorial focuses on Nim's macro system, which involves the execution of functions at compile-time to transform Nim syntax trees. The following examples illustrate the possibilities of macros:

- An assert macro that prints both sides of a comparison operator if the assertion fails. For instance, myAssert(a == b) is converted to if a != b: quit($a " != " $b).

- A debug macro that prints the value and the name of a symbol. For example, myDebugEcho(a) is converted to echo "a: ", a.

- Symbolic differentiation of an expression. For instance, diff(apow(x,3) + bpow(x,2) + cx + d, x) is converted to 3apow(x,2) + 2b*x + c.

Macro Arguments

Macro arguments can be approached from two perspectives: overload resolution and within the macro body. When invoking a macro like foo(arg: int) with an expression like foo(x), the variable x must be compatible with int for overload resolution. However, within the macro body, the parameter arg has the type NimNode, not int. This approach will become clearer with concrete examples.

There are two ways to pass arguments to a macro: through typed or untyped arguments.

Untyped Arguments

Untyped macro arguments are passed to the macro before undergoing semantic checks. The syntax tree passed to the macro doesn't need to make sense in terms of Nim yet; it only needs to be parsable. Typically, the macro uses the argument in the transformation result without performing checks on the argument itself. The compiler checks the result of a macro expansion, so major issues are unlikely to occur. However, untyped arguments don't interact well with Nim's overload resolution. On the positive side, the syntax tree is relatively predictable and less complex compared to typed arguments.

Typed Arguments

With typed arguments, the semantic checker processes the argument and applies transformations before passing it to the macro. Identifier nodes are resolved as symbols, implicit type conversions are represented as calls in the tree, templates are expanded, and nodes possess type information. Typed arguments can specify the type in the argument list, but other types like int, float, or MyObjectType are also considered typed arguments. They are passed to the macro as a syntax tree.

Static Arguments

Static arguments allow passing values as values, rather than syntax tree nodes, to a macro. For example, in the macro foo(arg: static[int]), the expression foo(x) requires x to be an integer constant. However, within the macro body, arg behaves like a regular parameter of type int.

```nim
import std/macros

macro myMacro(arg: static[int]): untyped =
  echo arg # just an int (7), not `NimNode`

myMacro(1 + 2 * 3)
```

Code Blocks as Arguments

A convenient approach is to pass the last argument of a call expression as a separate code block with proper indentation. This notation is useful for macros, as it enables the passing of syntax trees of varying complexity.

```nim
echo "Hello ":
  let a = "Wor"
  let b = "ld!"
  a & b
```

The Syntax Tree

Understanding how Nim source code is represented in a syntax tree and how to structure it is essential for constructing a Nim syntax tree. The macros module documents the nodes of the Nim syntax tree. Additionally, macros.treeRepr can convert a syntax tree into a multiline string suitable for printing on the console. It helps examine how argument expressions are represented in tree form, aiding in debugging and inspection. Another useful macro, dumpTree, prints its argument in tree representation without performing any other actions.

Nim Codes

```
dumpTree:
  var mt: MyType = MyType(a:123.456, b:"abcdef")

# output:
#   StmtList
#     VarSection
#       IdentDefs
#         Ident "mt"
```

```
#       Ident "MyType"
#       ObjConstr
#         Ident "MyType"
#         ExprColonExpr
#           Ident "a"
#           FloatLit 123.456
#         ExprColonExpr
#           Ident "b"
#           StrLit "abcdef"
```

Custom Semantic Checking

When working with macro arguments, it's important to verify whether the argument is in the appropriate form. While capturing every incorrect input may not be necessary, it's crucial to identify potential issues that could lead to a crash during macro evaluation and generate clear error messages. Macros provide tools like macros.expectKind and macros.expectLen for basic checks. For more complex checks, macros.error can be used to create custom error messages.

```
macro myAssert(arg: untyped): untyped =
  arg.expectKind nnkInfix
```

Generating Code

Approaches to Code Generation

There are two distinct approaches to code generation, each offering its own advantages. Let's delve into each approach in detail:

Approach 1: Constructing the Syntax Tree with newTree and newLit

The first approach involves constructing the syntax tree by utilizing expressions that incorporate multiple calls to newTree and newLit. This method provides granular control over syntax tree generation at a low level. By carefully constructing the tree using these calls, you can precisely manipulate and shape the code. This approach offers flexibility and customization options, making it suitable for complex code generation tasks. To streamline the process and reduce verbosity when using newTree and newLit calls, the macro macros.dumpAstGen can be employed.

Approach 2: Utilizing quote do: Expressions for Conciseness

The second approach leverages quote do: expressions for code generation. With this approach, you can directly write the code intended for generation, resulting in more concise and readable code. Backticks (`) are used to insert code from NimNode symbols into the generated expression, providing a convenient way to incorporate dynamic elements. This approach is ideal for simpler code generation tasks or situations where brevity is a priority. It abstracts away some of the low-level details of syntax tree construction, enabling faster and more straightforward code generation.

Important Considerations for Code Generation

When engaging in code generation, there are several crucial considerations and techniques to keep in mind:

Custom Prefix Operator and Accent Quoting:

- Define a custom prefix operator whenever backticks are needed for code injection.
- Accent quote the injected symbol when it resolves to a symbol.

Ensuring Type Safety:

- Only inject symbols of type NimNode into the generated syntax tree.
- Use newLit to convert arbitrary values into expression trees of type NimNode to guarantee type safety when injecting them into the tree.

Debugging and Inspecting Generated Code:

To debug and inspect the code generated by a macro, add the statement echo result.repr as the last line of the macro. This provides insights into the generated code and helps verify its correctness.

Building Your First Macro: Implementing myAssert

Let's now walk through implementing the myAssert macro. We'll start by creating a basic example of macro usage that simply prints the argument. This approach will help us understand the structure and requirements of a valid argument.

By examining the output, we observe that the argument represents an infix operator with a node kind of "Infix." Additionally, we can determine that the two operands are located at index 1 and 2. Armed with this knowledge, we can proceed to write the actual macro, incorporating the necessary logic to handle the argument structure.

For debugging and inspecting the generated code during development, you can utilize the echo result.repr statement, which has been used to obtain the aforementioned output.

Using Macros Effectively and Responsibly

Macros possess significant power, so it's crucial to use them wisely. Here are some important considerations:

- Use macros sparingly but effectively, as they can alter the semantics of expressions and potentially confuse others.
- If templates or generics can achieve the same logic without sacrificing clarity, prefer them over macros.
- Provide comprehensive documentation for macro usage to ensure clarity and understanding.

Limitations and Important Notes

Macros inherit the limitations of the NimVM compiler. Remember the following:

- Macros must be implemented using pure Nim code.
- Macros cannot directly call C functions, except those integrated into the compiler.
- While macros can initiate external processes on the shell, be mindful of these limitations.

By keeping these limitations and considerations in mind, you can harness the power of macros while maintaining code clarity and maintainability.

More Examples: Exploring the Possibilities

This tutorial covers the fundamentals of the macro system, but there are numerous macros available that can serve as sources of inspiration. Let's explore a few notable examples:

Strformat:

Within the Nim standard library, the strformat library includes a macro that performs string literal parsing at compile time. While parsing strings in macros is generally not recommended due to limitations and performance considerations, strformat showcases a practical use case for a slightly more complex macro. It extends beyond the simplicity of the assert macro and demonstrates the possibilities offered by macros in specific scenarios.

Source

Ast Pattern Matching:

The Ast Pattern Matching macro library is designed to facilitate the creation of intricate macros. It serves as an excellent example of how the Nim syntax tree can be leveraged with new semantics. This library demonstrates advanced techniques and showcases the power of macros in creating sophisticated code transformations.

Source

OpenGL Sandbox:

The OpenGL Sandbox project features a fully macro-based Nim to GLSL compiler. It utilizes macros extensively to scan and compile cross-library functions for execution on the GPU. This project showcases the ability of macros to enable complex code generation and integration with external systems.

By exploring these examples and others available in the Nim ecosystem, you can discover a wide range of possibilities and gain inspiration for your own macro-based solutions.

Source

Remember, macros offer great power, but it's important to use them judiciously and provide thorough documentation to ensure clarity and maintainability in your codebase.

Index

A

Advanced types 58
Annotating procs 122
Arrays..........................68
Assignment Statement 14
Ast Pattern Matching.154

B

Basic types 50
Bit fields 66
block statement 24
Booleans 50
Break statement 25

C

Case statement 16
Characters....................51
Code Blocks................142
Compilation 136
Constants.....................12
Continue statement.....26
Control flow statements
  15

Custom Semantic 144

D

Default values 38
Discard 36
Distinct Type 94
Dynamic dispatch 114

E

Enumerations 59
Exceptions..................118
Excluding symbols........99

F

First Program 5
Floats 55
For Statement..............20
Form statement 99
Forward declarations...43
Functions 46

G

Generating Code........146
Generics.....................124

H

hello world 5

I

If statement 15
Include statement 101
indentation 28
Inheritance 104
Integers 53
Iterators 47

J

JavaScript 136

L

Let Statement 13
Lexical Elements 8
Lifting Procs 135

M

Macro 149
Macro Arguments 139
Method call syntax 110
methods 46
Modules 95
Mutually recursive types
 106

N

Named arguments 37
Nim 64

O

Object Oriented
 Programming 103
Object variants 108
Objects 82
Open arrays 76
OpenGL Sandbox 155
Operators 42
Ordinal types 60
Overloaded procedures
 40

P

Parameters 34
Procedural type 93
Procedures 30
Programs 103
Properties 112

R

Raise statement 119
Reference 90

S

Sequences 73
Sets 63
Slices 79
Statements 28
Static Arguments 141
Strformat 154
Strings 52
Subranges 62
Syntax Tree 142

T

Templates 129
Try statement 120
Tuples 84
Type Conversion 56
Type conversions 107

U

Untyped Arguments .. 139

V

var statement 11
Varargs 78
variable 32

W

When statement 26
While Statement 19
writing macros 149

www.ingramcontent.com/pod-product-compliance
Lightning Source LLC
Chambersburg PA
CBHW071209240526
45470CB00018B/1651